This planner belongs to:

_____

Twenty-three

January

S	M	T	W	T	F	S
1	2	3	4	5	6	7
8	9	10	11	12	13	14
15	16	17	18	19	20	21
22	23	24	25	26	27	28
29	30	31				

February

S	M	T	W	T	F	S
			1	2	3	4
5	6	7	8	9	10	11
12	13	14	15	16	17	18
19	20	21	22	23	24	25
26	27	28				

March

S	M	T	W	T	F	S
			1	2	3	4
5	6	7	8	9	10	11
12	13	14	15	16	17	18
19	20	21	22	23	24	25
26	27	28	29	30	31	

April

S	M	T	W	T	F	S
						1
2	3	4	5	6	7	8
9	10	11	12	13	14	15
16	17	18	19	20	21	22
23	24	25	26	27	28	29
30						

May

S	M	T	W	T	F	S
	1	2	3	4	5	6
7	8	9	10	11	12	13
14	15	16	17	18	19	20
21	22	23	24	25	26	27
28	29	30	31			

June

S	M	T	W	T	F	S
				1	2	3
4	5	6	7	8	9	10
11	12	13	14	15	16	17
18	19	20	21	22	23	24
25	26	27	28	29	30	

July

S	M	T	W	T	F	S
						1
2	3	4	5	6	7	8
9	10	11	12	13	14	15
16	17	18	19	20	21	22
23	24	25	26	27	28	29
30	31					

August

S	M	T	W	T	F	S
		1	2	3	4	5
6	7	8	9	10	11	12
13	14	15	16	17	18	19
20	21	22	23	24	25	26
27	28	29	30	31		

September

S	M	T	W	T	F	S
					1	2
3	4	5	6	7	8	9
10	11	12	13	14	15	16
17	18	19	20	21	22	23
24	25	26	27	28	29	30

October

S	M	T	W	T	F	S
1	2	3	4	5	6	7
8	9	10	11	12	13	14
15	16	17	18	19	20	21
22	23	24	25	26	27	28
29	30	31				

November

S	M	T	W	T	F	S
			1	2	3	4
5	6	7	8	9	10	11
12	13	14	15	16	17	18
19	20	21	22	23	24	25
26	27	28	29	30		

December

S	M	T	W	T	F	S
					1	2
3	4	5	6	7	8	9
10	11	12	13	14	15	16
17	18	19	20	21	22	23
24	25	26	27	28	29	30
31						

Year in Pixels

	J	F	M	A	M	J	J	A	S	O	N	D
1.												
2.												
3.												
4.												
5.												
6.												
7.												
8.												
9.												
10.												
11.												
12.												
13.												
14.												
15.												
16.												
17.												
18.												
19.												
20.												
21.												
22.												
23.												
24.												
25.												
26.												
27.												
28.												
29.												
30.												
31.												

Color Codes

Notes

January

MONDAY	TUESDAY	WEDNESDAY	THURSDAY
2	3	4	5
9	10	11	12
16	17	18	19
23	24	25	26

January

FRIDAY	SATURDAY	SUNDAY	NOTES
		1	○
			○
			○
			○
			○
6	7	8	○
			○
			○
			○
13	14	15	○
			○
			○
			○
			○
20	21	22	○
			○
			○
			○
			○
27	28	29	30
			31

February

MONDAY	TUESDAY	WEDNESDAY	THURSDAY
		1	2
6	7	8	9
13	14	15	16
20	21	22	23
27	28		

February

2023

FRIDAY	SATURDAY	SUNDAY	NOTES
3	4	5	○
			○
			○
			○
			○
10	11	12	○
			○
			○
			○
			○
17	18	19	○
			○
			○
			○
			○
24	25	26	○
			○
			○
			○
			○
			NOTES

March

MONDAY	TUESDAY	WEDNESDAY	THURSDAY
		1	2
6	7	8	9
13	14	15	16
20	21	22	23
27	28	29	30

March

FRIDAY	SATURDAY	SUNDAY	NOTES
3	4	5	○
			○
			○
			○
			○
10	11	12	○
			○
			○
			○
17	18	19	○
			○
			○
			○
			○
24	25	26	○
			○
			○
			○
			○
31			NOTES

April

MONDAY	TUESDAY	WEDNESDAY	THURSDAY
3	4	5	6
10	11	12	13
17	18	19	20
24	25	26	27

April

FRIDAY	SATURDAY	SUNDAY	NOTES
	1	2	○
			○
			○
			○
			○
7	8	9	○
			○
			○
			○
			○
14	15	16	○
			○
			○
			○
			○
21	22	23	○
			○
			○
			○
			○
28	29	30	NOTES

May

2023

MONDAY	TUESDAY	WEDNESDAY	THURSDAY
1	2	3	4
8	9	10	11
15	16	17	18
22	23	24	25
29	30	31	

May

FRIDAY	SATURDAY	SUNDAY	NOTES
5	6	7	○
			○
			○
			○
			○
12	13	14	○
			○
			○
			○
			○
19	20	21	○
			○
			○
			○
			○
26	27	28	○
			○
			○
			○
			○
			NOTES

June

MONDAY	TUESDAY	WEDNESDAY	THURSDAY
			1
5	6	7	8
12	13	14	15
19	20	21	22
26	27	28	29

June

2023

FRIDAY	SATURDAY	SUNDAY	NOTES
2	3	4	○
			○
			○
			○
			○
9	10	11	○
			○
			○
			○
16	17	18	○
			○
			○
			○
			○
23	24	25	○
			○
			○
			○
			○
30			NOTES

July

MONDAY	TUESDAY	WEDNESDAY	THURSDAY
3	4	5	6
10	11	12	13
17	18	19	20
24	25	26	27

July

FRIDAY	SATURDAY	SUNDAY	NOTES
	1	2	○
			○
			○
			○
			○
7	8	9	○
			○
			○
			○
14	15	16	○
			○
			○
			○
			○
21	22	23	○
			○
			○
			○
			○
28	29	30	31

August

MONDAY	TUESDAY	WEDNESDAY	THURSDAY
	1	2	3
7	8	9	10
14	15	16	17
21	22	23	24
28	29	30	31

August

2023

FRIDAY	SATURDAY	SUNDAY	NOTES
4	5	6	○
			○
			○
			○
			○
11	12	13	○
			○
			○
			○
18	19	20	○
			○
			○
			○
			○
25	26	27	NOTES

September 2023

MONDAY	TUESDAY	WEDNESDAY	THURSDAY
4	5	6	7
11	12	13	14
18	19	20	21
25	26	27	28

September
2023

FRIDAY	SATURDAY	SUNDAY	NOTES
1	2	3	○
			○
			○
			○
			○
8	9	10	○
			○
			○
			○
15	16	17	○
			○
			○
			○
			○
22	23	24	○
			○
			○
			○
			○
29	30		NOTES

October  2023

MONDAY	TUESDAY	WEDNESDAY	THURSDAY
2	3	4	5
9	10	11	12
16	17	18	19
23	24	25	26

October

2023

FRIDAY	SATURDAY	SUNDAY	NOTES
		1	○
			○
			○
			○
			○
6	7	8	○
			○
			○
			○
13	14	15	○
			○
			○
			○
			○
20	21	22	○
			○
			○
			○
			○
27	28	29	30 / 31

November 2023

MONDAY	TUESDAY	WEDNESDAY	THURSDAY
		1	2
6	7	8	9
13	14	15	16
20	21	22	23
27	28	29	30

November

FRIDAY	SATURDAY	SUNDAY	NOTES
3	4	5	○
			○
			○
			○
			○
10	11	12	○
			○
			○
			○
			○
17	18	19	○
			○
			○
			○
			○
24	25	26	○
			○
			○
			○
			○
			NOTES

December 2023

MONDAY	TUESDAY	WEDNESDAY	THURSDAY
4	5	6	7
11	12	13	14
18	19	20	21
25	26	27	28

December

FRIDAY	SATURDAY	SUNDAY	NOTES
1	2	3	○
			○
			○
			○
			○
8	9	10	○
			○
			○
			○
			○
15	16	17	○
			○
			○
			○
			○
22	23	24	○
			○
			○
			○
			○
29	30	31	NOTES

December
2022

01 THURSDAY

02 FRIDAY

03 SATURDAY

04 SUNDAY

December
2022

05 MONDAY

06 TUESDAY

07 WEDNESDAY

08 THURSDAY

09 FRIDAY

10 SATURDAY

11 SUNDAY

12 MONDAY

December
2022

13 TUESDAY

14 WEDNESDAY

15 THURSDAY

16 FRIDAY

December
2022

17 SATURDAY

○
○
○
○
○
○
○
○

18 SUNDAY

○
○
○
○
○
○
○
○

19 MONDAY

○
○
○
○
○
○
○
○

20 TUESDAY

○
○
○
○
○

December
2022

21 WEDNESDAY

○
○
○
○
○
○
○
○

22 THURSDAY

○
○
○
○
○
○
○
○

23 FRIDAY

○
○
○
○
○
○
○
○

24 SATURDAY

○
○
○
○
○

December
2022

25 SUNDAY

26 MONDAY

27 TUESDAY

28 WEDNESDAY

December
2022

29 THURSDAY

○
○
○
○
○
○
○
○

30 FRIDAY

○
○
○
○
○
○
○
○

31 SATURDAY

○
○
○
○
○
○
○
○

NOTES

January 2023

01 SUNDAY

- ○
- ○
- ○
- ○
- ○
- ○
- ○
- ○

02 MONDAY

- ○
- ○
- ○
- ○
- ○
- ○
- ○
- ○

03 TUESDAY

- ○
- ○
- ○
- ○
- ○
- ○
- ○
- ○

04 WEDNESDAY

- ○
- ○
- ○
- ○

05 THURSDAY

06 FRIDAY

07 SATURDAY

08 SUNDAY

09 MONDAY
○
○
○
○
○
○
○

10 TUESDAY
○
○
○
○
○
○
○
○

11 WEDNESDAY
○
○
○
○
○
○
○
○

12 THURSDAY
○
○
○
○
○

13 FRIDAY

14 SATURDAY

15 SUNDAY

16 MONDAY

17 TUESDAY

○
○
○
○
○
○
○

18 WEDNESDAY

○
○
○
○
○
○
○
○

19 THURSDAY

○
○
○
○
○
○
○
○

20 FRIDAY

○
○
○
○
○

21 SATURDAY

22 SUNDAY

23 MONDAY

24 TUESDAY

25 WEDNESDAY

○
○
○
○
○
○
○
○

26 THURSDAY

○
○
○
○
○
○
○
○

27 FRIDAY

○
○
○
○
○
○
○
○

28 SATURDAY

○
○
○
○
○

January 2023

29 SUNDAY

- ○
- ○
- ○
- ○
- ○
- ○
- ○
- ○

30 MONDAY

- ○
- ○
- ○
- ○
- ○
- ○
- ○
- ○

31 TUESDAY

- ○
- ○
- ○
- ○
- ○
- ○
- ○
- ○

NOTES

February 2023

01 WEDNESDAY

○
○
○
○
○
○
○
○

02 THURSDAY

○
○
○
○
○
○
○
○

03 FRIDAY

○
○
○
○
○
○
○
○

04 SATURDAY

○
○
○
○
○

February 2023

05 SUNDAY

06 MONDAY

07 TUESDAY

08 WEDNESDAY

February 2023

09 THURSDAY

○
○
○
○
○
○
○
○

10 FRIDAY

○
○
○
○
○
○
○
○

11 SATURDAY

○
○
○
○
○
○
○
○

12 SUNDAY

○
○
○
○
○

13 MONDAY

14 TUESDAY

15 WEDNESDAY

16 THURSDAY

17 FRIDAY

18 SATURDAY

19 SUNDAY

20 MONDAY

February 2023

21 TUESDAY

22 WEDNESDAY

23 THURSDAY

24 FRIDAY

25 SATURDAY

26 SUNDAY

27 MONDAY

28 TUESDAY

March 2023

01 WEDNESDAY

02 THURSDAY

03 FRIDAY

04 SATURDAY

March
2023

05 SUNDAY

06 MONDAY

07 TUESDAY

08 WEDNESDAY

09 THURSDAY

10 FRIDAY

11 SATURDAY

12 SUNDAY

13 MONDAY

14 TUESDAY

15 WEDNESDAY

16 THURSDAY

March
2023

17 FRIDAY

18 SATURDAY

19 SUNDAY

20 MONDAY

21 TUESDAY

22 WEDNESDAY

23 THURSDAY

24 FRIDAY

25 SATURDAY

26 SUNDAY

27 MONDAY

28 TUESDAY

29 WEDNESDAY

○
○
○
○
○
○
○
○

30 THURSDAY

○
○
○
○
○
○
○
○

31 FRIDAY

○
○
○
○
○
○
○
○

NOTES

01 SATURDAY

02 SUNDAY

03 MONDAY

04 TUESDAY

April
2023

05 WEDNESDAY

○
○
○
○
○
○
○

06 THURSDAY

○
○
○
○
○
○
○
○

07 FRIDAY

○
○
○
○
○
○
○
○

08 SATURDAY

○
○
○
○
○

09 SUNDAY

10 MONDAY

11 TUESDAY

12 WEDNESDAY

13 THURSDAY

14 FRIDAY

15 SATURDAY

16 SUNDAY

17 MONDAY

18 TUESDAY

19 WEDNESDAY

20 THURSDAY

April 2023

21 FRIDAY

22 SATURDAY

23 SUNDAY

24 MONDAY

April 2023

25 TUESDAY

○ ──────────────────────────────────
○ ──────────────────────────────────
○ ──────────────────────────────────
○ ──────────────────────────────────
○ ──────────────────────────────────
○ ──────────────────────────────────
○ ──────────────────────────────────
○ ──────────────────────────────────

26 WEDNESDAY

○ ──────────────────────────────────
○ ──────────────────────────────────
○ ──────────────────────────────────
○ ──────────────────────────────────
○ ──────────────────────────────────
○ ──────────────────────────────────
○ ──────────────────────────────────
○ ──────────────────────────────────

27 THURSDAY

○ ──────────────────────────────────
○ ──────────────────────────────────
○ ──────────────────────────────────
○ ──────────────────────────────────
○ ──────────────────────────────────
○ ──────────────────────────────────
○ ──────────────────────────────────
○ ──────────────────────────────────

28 FRIDAY

○ ──────────────────────────────────
○ ──────────────────────────────────
○ ──────────────────────────────────
○ ──────────────────────────────────
○ ──────────────────────────────────

April 2023

29 SATURDAY
○
○
○
○
○
○
○
○

30 SUNDAY
○
○
○
○
○
○
○
○

NOTES

01 MONDAY

02 TUESDAY

03 WEDNESDAY

04 THURSDAY

May 2023

05 FRIDAY

○
○
○
○
○
○
○
○

06 SATURDAY

○
○
○
○
○
○
○
○

07 SUNDAY

○
○
○
○
○
○
○
○

08 MONDAY

○
○
○
○
○

09 TUESDAY

10 WEDNESDAY

11 THURSDAY

12 FRIDAY

May
2023

13 SATURDAY

14 SUNDAY

15 MONDAY

16 TUESDAY

17 WEDNESDAY

18 THURSDAY

19 FRIDAY

20 SATURDAY

May
2023

21 SUNDAY

○
○
○
○
○
○
○
○

22 MONDAY

○
○
○
○
○
○
○
○

23 TUESDAY

○
○
○
○
○
○
○
○

24 WEDNESDAY

○
○
○
○

25 THURSDAY

26 FRIDAY

27 SATURDAY

28 SUNDAY

May
2023

29 MONDAY
○
○
○
○
○
○
○
○

30 TUESDAY
○
○
○
○
○
○
○
○

31 WEDNESDAY
○
○
○
○
○
○
○
○

NOTES

June 2023

01 THURSDAY

02 FRIDAY

03 SATURDAY

04 SUNDAY

05 MONDAY

06 TUESDAY

07 WEDNESDAY

08 THURSDAY

June 2023

09 FRIDAY

10 SATURDAY

11 SUNDAY

12 MONDAY

13 TUESDAY

○
○
○
○
○
○
○
○

14 WEDNESDAY

○
○
○
○
○
○
○
○

15 THURSDAY

○
○
○
○
○
○
○
○

16 FRIDAY

○
○
○
○
○

17 SATURDAY

18 SUNDAY

19 MONDAY

20 TUESDAY

June 2023

21 WEDNESDAY

○
○
○
○
○
○
○
○

22 THURSDAY

○
○
○
○
○
○
○
○

23 FRIDAY

○
○
○
○
○
○
○
○

24 SATURDAY

○
○
○
○
○

25 SUNDAY

26 MONDAY

27 TUESDAY

28 WEDNESDAY

29 THURSDAY

○ —————————————————————————————
○ —————————————————————————————
○ —————————————————————————————
○ —————————————————————————————
○ —————————————————————————————
○ —————————————————————————————
○ —————————————————————————————
○ —————————————————————————————

30 FRIDAY

○ —————————————————————————————
○ —————————————————————————————
○ —————————————————————————————
○ —————————————————————————————
○ —————————————————————————————
○ —————————————————————————————
○ —————————————————————————————
○ —————————————————————————————

NOTES

July
2023

01 SATURDAY

02 SUNDAY

03 MONDAY

04 TUESDAY

July

2023

05 WEDNESDAY

06 THURSDAY

07 FRIDAY

08 SATURDAY

09 SUNDAY

10 MONDAY

11 TUESDAY

12 WEDNESDAY

13 THURSDAY

14 FRIDAY

15 SATURDAY

16 SUNDAY

17 MONDAY

○
○
○
○
○
○
○
○

18 TUESDAY

○
○
○
○
○
○
○
○
○

19 WEDNESDAY

○
○
○
○
○
○
○
○

20 THURSDAY

○
○
○
○
○

July
2023

21 FRIDAY
○
○
○
○
○
○
○
○

22 SATURDAY
○
○
○
○
○
○
○
○

23 SUNDAY
○
○
○
○
○
○
○
○

24 MONDAY
○
○
○
○
○

July
2023

25 TUESDAY

26 WEDNESDAY

27 THURSDAY

28 FRIDAY

July
2023

29 SATURDAY

30 SUNDAY

31 MONDAY

NOTES

01 TUESDAY

02 WEDNESDAY

03 THURSDAY

04 FRIDAY

August
2023

05 SATURDAY

06 SUNDAY

07 MONDAY

08 TUESDAY

09 WEDNESDAY

10 THURSDAY

11 FRIDAY

12 SATURDAY

13 SUNDAY

14 MONDAY

15 TUESDAY

16 WEDNESDAY

17 THURSDAY

18 FRIDAY

19 SATURDAY

20 SUNDAY

August
2023

21 MONDAY

○
○
○
○
○
○
○
○

22 TUESDAY

○
○
○
○
○
○
○
○

23 WEDNESDAY

○
○
○
○
○
○
○
○

24 THURSDAY

○
○
○
○
○

August 2023

25 FRIDAY

26 SATURDAY

27 SUNDAY

28 MONDAY

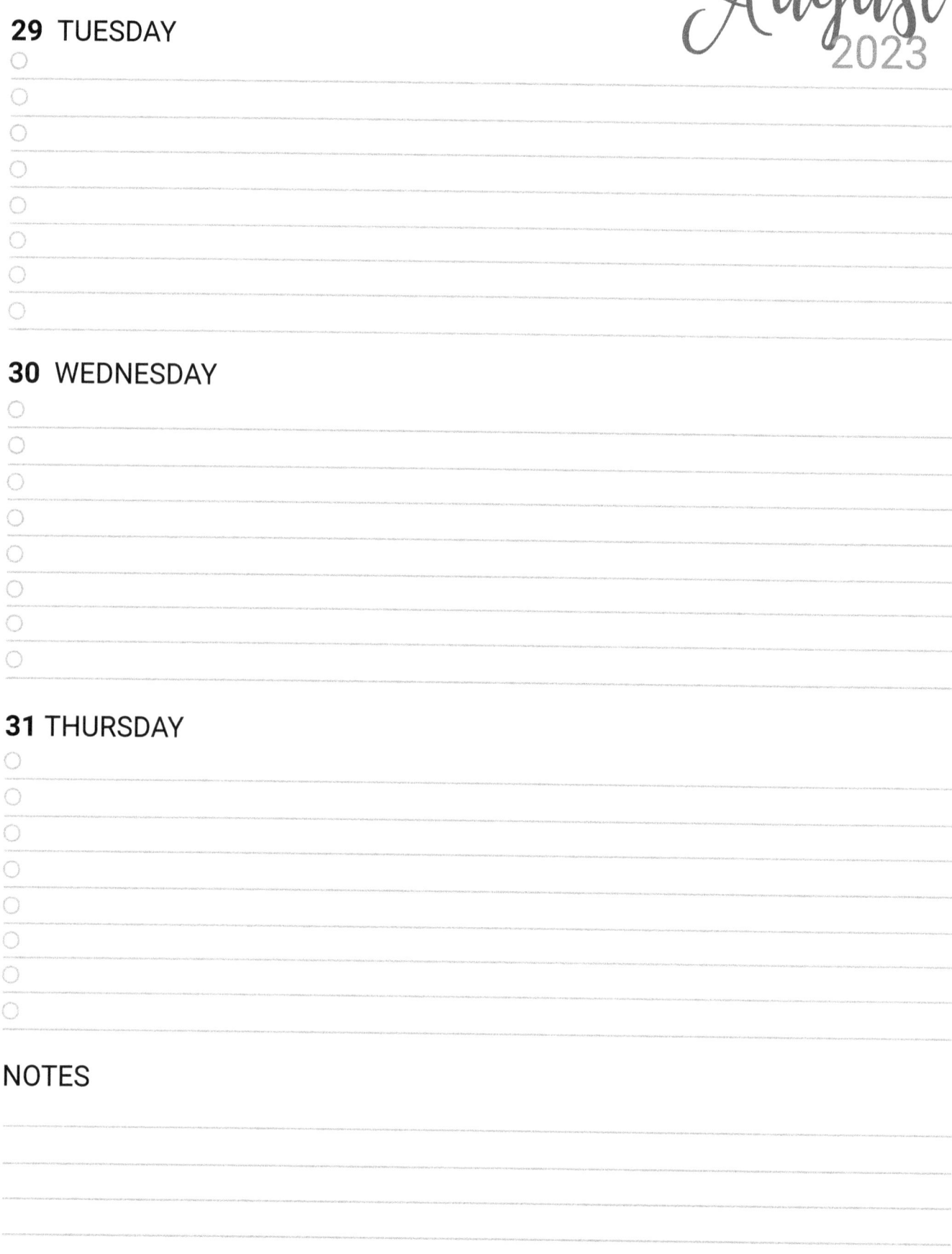

August 2023

29 TUESDAY

○
○
○
○
○
○
○
○

30 WEDNESDAY

○
○
○
○
○
○
○
○

31 THURSDAY

○
○
○
○
○
○
○
○

NOTES

01 FRIDAY

02 SATURDAY

03 SUNDAY

04 MONDAY

05 TUESDAY

06 WEDNESDAY

07 THURSDAY

08 FRIDAY

09 SATURDAY

○
○
○
○
○
○
○
○

10 SUNDAY

○
○
○
○
○
○
○
○

11 MONDAY

○
○
○
○
○
○
○
○

12 TUESDAY

○
○
○
○
○

13 WEDNESDAY

14 THURSDAY

15 FRIDAY

16 SATURDAY

17 SUNDAY

○
○
○
○
○
○
○
○

18 MONDAY

○
○
○
○
○
○
○
○
○

19 TUESDAY

○
○
○
○
○
○
○
○

20 WEDNESDAY

○
○
○
○
○

September
2023

21 THURSDAY

22 FRIDAY

23 SATURDAY

24 SUNDAY

September

2023

25 MONDAY

26 TUESDAY

27 WEDNESDAY

28 THURSDAY

29 FRIDAY

○ _____
○ _____
○ _____
○ _____
○ _____
○ _____
○ _____
○ _____

30 SATURDAY

○ _____
○ _____
○ _____
○ _____
○ _____
○ _____
○ _____
○ _____

NOTES

_____
_____
_____
_____
_____
_____
_____
_____
_____
_____
_____
_____

October
2023

01 SUNDAY

02 MONDAY

03 TUESDAY

04 WEDNESDAY

05 THURSDAY

06 FRIDAY

07 SATURDAY

08 SUNDAY

October 2023

09 MONDAY

10 TUESDAY

11 WEDNESDAY

12 THURSDAY

13 FRIDAY

14 SATURDAY

15 SUNDAY

16 MONDAY

17 TUESDAY

18 WEDNESDAY

19 THURSDAY

20 FRIDAY

21 SATURDAY

22 SUNDAY

23 MONDAY

24 TUESDAY

October
2023

25 WEDNESDAY

26 THURSDAY

27 FRIDAY

28 SATURDAY

October
2023

29 SUNDAY

30 MONDAY

31 TUESDAY

NOTES

November
2023

01 WEDNESDAY

02 THURSDAY

03 FRIDAY

04 SATURDAY

November
2023

05 SUNDAY

○
○
○
○
○
○
○
○

06 MONDAY

○
○
○
○
○
○
○
○

07 TUESDAY

○
○
○
○
○
○
○
○

08 WEDNESDAY

○
○
○
○

November
2023

09 THURSDAY

10 FRIDAY

11 SATURDAY

12 SUNDAY

November
2023

13 MONDAY

○
○
○
○
○
○
○
○

14 TUESDAY

○
○
○
○
○
○
○
○

15 WEDNESDAY

○
○
○
○
○
○
○
○

16 THURSDAY

○
○
○
○
○

November
2023

17 FRIDAY

18 SATURDAY

19 SUNDAY

20 MONDAY

21 TUESDAY

○
○
○
○
○
○
○
○

22 WEDNESDAY

○
○
○
○
○
○
○
○

23 THURSDAY

○
○
○
○
○
○
○
○

24 FRIDAY

○
○
○
○
○

November
2023

25 SATURDAY

26 SUNDAY

27 MONDAY

28 TUESDAY

29 WEDNESDAY

- ○
- ○
- ○
- ○
- ○
- ○
- ○
- ○

30 THURSDAY

- ○
- ○
- ○
- ○
- ○
- ○
- ○
- ○

NOTES

December
2023

01 FRIDAY

02 SATURDAY

03 SUNDAY

04 MONDAY

December
2023

05 TUESDAY

○
○
○
○
○
○
○

06 WEDNESDAY

○
○
○
○
○
○
○

07 THURSDAY

○
○
○
○
○
○
○

08 FRIDAY

○
○
○
○

09 SATURDAY

○
○
○
○
○
○
○
○

10 SUNDAY

○
○
○
○
○
○
○
○

11 MONDAY

○
○
○
○
○
○
○
○

12 TUESDAY

○
○
○
○
○

December
2023

13 WEDNESDAY

14 THURSDAY

15 FRIDAY

16 SATURDAY

December
2023

17 SUNDAY

18 MONDAY

19 TUESDAY

20 WEDNESDAY

December
2023

21 THURSDAY

○
○
○
○
○
○
○
○

22 FRIDAY

○
○
○
○
○
○
○
○

23 SATURDAY

○
○
○
○
○
○
○
○

24 SUNDAY

○
○
○
○
○

25 MONDAY

26 TUESDAY

27 WEDNESDAY

28 THURSDAY

December
2023

29 FRIDAY

○
○
○
○
○
○
○
○

30 SATURDAY

○
○
○
○
○
○
○
○

31 SUNDAY

○
○
○
○
○
○
○
○

NOTES

www.ingramcontent.com/pod-product-compliance
Lightning Source LLC
Chambersburg PA
CBHW081333120626
46546CB00011B/3335